Joseph Kibange Mihali

Lake Tanganyika in peril

Joseph Kibange Mihali

Lake Tanganyika in peril

Understanding and responding to rising sea levels

ScienciaScripts

Imprint

Any brand names and product names mentioned in this book are subject to trademark, brand or patent protection and are trademarks or registered trademarks of their respective holders. The use of brand names, product names, common names, trade names, product descriptions etc. even without a particular marking in this work is in no way to be construed to mean that such names may be regarded as unrestricted in respect of trademark and brand protection legislation and could thus be used by anyone.

Cover image: www.ingimage.com

This book is a translation from the original published under ISBN 978-620-6-71845-1.

Publisher:
Sciencia Scripts
is a trademark of
Dodo Books Indian Ocean Ltd. and OmniScriptum S.R.L publishing group

120 High Road, East Finchley, London, N2 9ED, United Kingdom
Str. Armeneasca 28/1, office 1, Chisinau MD-2012, Republic of Moldova, Europe
Printed at: see last page
ISBN: 978-620-7-94831-4

Lake Tanganyika in Peril

Understanding and taking action against rising water levels

A word from the author
I'm Joseph KIBAnge MIHALI, a geological engineer by training, Founder and Managing Director of SCITRA SARL, an innovative company specializing in converting waste into finished products. We transform plastic waste into ecological paving stones, biodegradable waste into ecological charcoal and organic fertilizers. Our mission is twofold: to protect the environment while creating economic opportunities for local communities. As Head of Organizational Development for the Red Cross Urban Committee of the Democratic Republic of Congo (DRC) in Uvira, I have seen first-hand the environmental and socio-economic challenges facing our communities. The rising waters of Lake Tanganyika are one of the most urgent and complex problems we face.

The rising waters of Lake Tanganyika are a phenomenon with many and varied consequences. It affects not only aquatic and terrestrial ecosystems, but also the daily lives of local populations. This book is the fruit of many years of observation, research and involvement in the field. My aim is to provide an in-depth understanding of this phenomenon, its causes, impacts and possible solutions.

The environmental and socio-economic challenges posed by the rising waters of Lake Tanganyika are numerous. Riparian communities are often the first and hardest hit by flooding and forced displacement. Traditional lifestyles and economic activities, such as fishing and farming, are profoundly disrupted. As Head of Organizational Development at the Red Cross OF THE DEMOCRATIC REPUBLIC OF CONGO, I have worked on several projects aimed at mitigating these impacts, notably through community resilience and disaster risk management initiatives.

The book also explores local and international initiatives to manage Lake Tanganyika's resources sustainably. Cooperation between the countries bordering the lake is essential to meet this cross-border challenge. Organizations such as the Red Cross, LATAWAMA and AFPDE play a crucial role in implementing conservation and resource management projects. Their work on the ground, combined with emission reduction and resource management policies, is an integral part of the response to rising waters.

As founder and managing director of SCITRA SARL, I have also sought to integrate innovative environmental solutions into our operations. Turning waste into finished products is not just a question of waste management; it's also a strategy for adapting to environmental change. By transforming plastic waste into eco-friendly paving stones and biodegradable waste into eco-friendly charcoal and organic fertilizer, we help to reduce pressure on natural resources and create sustainable livelihoods for local communities.

Writing this book has been an enriching and often moving journey.I've had the opportunity to meet and work with some extraordinary people, whose resilience and commitment are a constant source of inspiration.I'd like to thank all the people and organizations who contributed to the making of this book. Their knowledge, support and collaboration have been invaluable.

I hope this book will provide valuable information and help raise awareness of the critical issues surrounding the rising waters of Lake Tanganyika. More importantly, I hope it will encourage concrete action for the conservation and sustainable management of this vital lake. Together, we can make a difference. Let's continue to work hand in hand to preserve Lake Tanganyika and its ecosystems, while ensuring a sustainable and prosperous future for our communities.

Joseph KIBANGE MIHALI
Geological engineer
Founder and Managing Director of SCITRA SARL
Head of Organizational Development
Red Cross RDC, uvira

Table of Contents

Chapter 1. Introduction

Introduction

Lake Tanganyika, the second largest freshwater lake in the world, is facing a growing threat from rising water levels. This rapid and worrying rise in lake levels is endangering local populations, their economic activities and the fragile environment of this Central African region.

Since the 2010s, scientific observations have revealed a substantial rise in the water level of Lake Tanganyika, with already visible and worrying consequences. The shores have gradually flooded, forcing many communities to relocate. Fishing and farming, the mainstays of the local economy, have been seriously disrupted. The lake's unique biodiversity, a UNESCO World Heritage Site, is also threatened by these brutal changes.

Faced with this alarming situation, it is essential to understand the causes of rising water levels and to identify concrete solutions. The aim of this book is to provide a detailed analysis of the factors behind this phenomenon, whether climatic, hydrological or linked to human activities. It will also propose a set of adaptation and mitigation measures, based on international best practice and adapted to the local context.

Through a multidisciplinary approach involving scientific experts, political decision-makers and local communities, this book aims to provide a pragmatic and ambitious framework for action to meet the challenge of rising water levels in Lake Tanganyika. This is a major challenge for preserving fragile ecosystems, ensuring the safety and resilience of local populations, and guaranteeing the sustainable development of this region.

1.1. Geographical and ecological context of Lake Tanganyika

Lake Tanganyika, located in Africa's Great Lakes region, is a non

not only one of the largest freshwater lakes in the world, but also one of the oldest and deepest. This chapter will explore in depth its precise geographical location, its climate and natural environment, as well as its key ecosystems and habitats.

1.2. Geographical location

Lake Tanganyika covers a surface area of almost 32,900 kmq, crossing the following countries: the Democratic Republic of Congo (DRC), Burundi, Tanzania and Zambia. It lies mainly in the Great African Rift Valley, a geologically active and diverse region.

1.2.1. **Geographical distribution**

The lake runs roughly north-south, with most of its length between latitudes 3° and 9° south. This geographical position influences not only the climate, but also the biodiversity of the lake and its surroundings.

1.2.2. **Geographical features**

The relief around Lake Tanganyika is marked by steep mountains and deep valleys, creating a spectacular and diverse landscape. The lake's coasts vary according to the underlying geology, from sandy beaches to sheer cliffs.

1.3 Climate and natural environment

The climate around Lake Tanganyika is influenced by its equatorial geographical position and the topographical features of the rift region. This section will explore the climatic aspects as well as the characteristics of the natural environment surrounding the lake.

1.3.1. **Climate**

The climate is generally tropical, with well-defined dry and wet seasons. Rainfall varies considerably along the lake, influencing vegetation and the reproductive cycles of aquatic species.

1.3.2. **Natural environment**

The surrounding areas include a combination of tropical rainforests, savannahs and farmland. These ecosystems play a crucial role in regulating the local climate and conserving biodiversity.

main ecosystems and habitats

Lake Tanganyika is renowned for its exceptional biodiversity, harboring a multitude of endemic species and offering varied habitats for aquatic fauna and flora. This section looks in detail at the main ecosystems and habitats in and around the lake.

aquatic ecosystems

Lake Tanganyika is divided into different depth zones, each home to species adapted to specific conditions. Shallow littoral zones are rich in biodiversity, while deeper waters are home to species adapted to high hydrostatic pressures.

Terrestrial habitats

The terrestrial areas adjacent to the lake include a diversity of habitats, from mountain forests to floodplains and agricultural areas. These habitats support a variety of terrestrial species that depend directly on the resources provided by Lake Tanganyika.

By exploring these aspects in depth, we can better appreciate the complexity and ecological importance of Lake Tanganyika, as well as the environmental challenges it faces today.

importance of the lake for local communities and biodiversity

Lake Tanganyika is central to the lives of local communities and is of inestimable ecological importance. Its wealth of natural resources and unique biodiversity make it an ecosystem of exceptional value to both mankind and nature. This section explores in depth the use of the lake by local populations and its ecological importance.

Use of the lake by local populations

Water

Lake Tanganyika is a vital source of freshwater for local communities. As the world's second largest freshwater lake by volume, it provides an essential resource for domestic, agricultural and industrial uses.

Domestic use

- The local populations depend heavily on the lake for their daily drinking water needs. Due to the absence of sophisticated water treatment infrastructures in many regions, lake water is often used directly after simple filtration.

- Local and international initiatives are working to improve water quality and provide sustainable solutions for access to drinking water, by implementing purification projects and water distribution systems.

Agricultural use

- Irrigation is a common practice in the agricultural areas around the lake. Water from the lake is used to irrigate crops, particularly during periods of drought.

- Irrigated agriculture from the lake helps to sustain the livelihoods of local farmers and guarantee food security in the region.

Industrial use

- In some areas, the lake's water is used for industrial processes, notably in small local enterprises processing agricultural and fishery products, and in the cement industry at Kabimba.

- Although industrial use of lake water is limited compared to other uses, it nevertheless contributes to the local economy.

Fishing

Fishing is one of the main economic and cultural activities around Lake Tanganyika. The lake is home to a wide variety of fish, including many endemic species, making it an indispensable source of livelihood for the local populations.

Artisanal fishing

- Artisanal fishing is practiced by thousands of local fishermen, who use traditional techniques to catch fish. This activity is often family-run, passing on ancestral know-how.

- The most commonly caught species include Lake Tanganyika sardines (Limnothrissa miodon and Stolothrissa tanganicae) and several cichlid species, which are both consumed locally and sold on regional markets.

Commercial fishing

- Commercial fishing, although less widespread than artisanal fishing, also plays an important role. Larger companies use
motorized boats and large nets for larger catches.

- This larger-scale fishing can pose sustainability challenges, requiring rigorous management to avoid overfishing and ensure the sustainability of fish stocks.

Fish processing and trading

- Fish caught in the lake is often smoked, dried or salted to preserve it. These traditional methods extend the shelf life of fish products and enable them to be marketed in more remote regions.

- The fish trade is an important source of income for local communities, making a significant contribution to the regional economy.

Transport

Lake Tanganyika is also a crucial transport route for local populations. Its length and geographical position make it a natural route for trade and travel.

Freight transport

- Boats and pirogues are commonly used to transport goods between riverside villages and the main towns along the lake. These include agricultural produce, fish and other consumer goods.

- Transport by water is often faster and less costly than by road, especially in regions where road infrastructure is limited or in poor condition.

Passenger transport

- Local people regularly use the lake to get around. Ferry services and private boats facilitate travel between the various communities along the lake.

- Lake transport plays a crucial role in the social and economic integration of lakeside regions, providing access to markets, health services and educational establishments.

Ecological importance

Lake Tanganyika is of exceptional ecological importance, harboring unique biological diversity and playing a crucial role in regional and global ecosystems.

Endemic species

Lake Tanganyika is famous for its endemic flora and fauna. Due to its geological age and relative isolation, many species have evolved in unique ways, creating a rich and diverse ecosystem.

Cichlids

- The lake is home to around 250 species of cichlids, over β8% of which are endemic. These fish are renowned for their bright colors and complex behaviors.

- Lake Tanganyika cichlids have evolved to occupy a wide variety of ecological niches, from predators to herbivores, and from rock-dwelling to pelagic species.

Other fish

- In addition to cichlids, the lake is also home to other endemic fish species, such as Lake Tanganyika sardines and several species of catfish.

- These fish play a crucial role in the lake's food webs, contributing to the overall health of the ecosystem.

Invertebrates and aquatic plants

- Lake Tanganyika is also rich in endemic invertebrates, including snails, shrimps and freshwater sponges.

- Endemic aquatic plants, such as water lily and algae species, provide essential habitats for many other species and contribute to the lake's primary productivity.

Biodiversity

Lake Tanganyika's biodiversity is not only impressive in terms of species numbers, but also plays a key role in the stability and resilience of adjacent aquatic and terrestrial ecosystems.

Ecological interactions

- Interactions between species in Lake Tanganyika are complex and well-balanced. Fish, invertebrates and aquatic plants form dynamic food webs that support the productivity and health of the ecosystem.

- For example, predatory cichlids keep smaller fish populations in check, while herbivores keep aquatic plant growth under control.

Adjacent terrestrial ecosystems

- The terrestrial ecosystems around the lake, such as riparian forests and wetlands, are closely linked to the health of the lake. These habitats provide nutrients and shelter for many aquatic and terrestrial species.

- Terrestrial biodiversity includes mammals, birds, reptiles and amphibians, many of which depend on the lake's resources for their survival.

Role in the carbon cycle

- Lake Tanganyika plays an important role in the regional carbon cycle. Biological processes, such as photosynthesis by aquatic plants and respiration by organisms, influence carbon flows between water, atmosphere and sediments.

- The lake's sediments also store large quantities of organic carbon, helping to regulate the global climate.

In conclusion, Lake Tanganyika is a crucially important ecosystem for local communities and global biodiversity. Its multifunctional role as a source of water, a fishing ground and a transport route is indispensable for the populations living along its shores. Moreover, its ecological richness, characterized by an exceptional diversity of endemic species and ecological complexity, underlines the need to preserve and protect this unique lake. Sustainable management and conservation of Lake Tanganyika are essential to ensure that its resources and ecological services continue to benefit present and future generations.

Rising sea levels: causes and consequences
A look at the challenges of rising sea levels
The rising waters of Lake Tanganyika represent a major environmental and

socio-economic challenge for the region. Due to various anthropogenic and natural factors, the lake level has risen significantly in recent decades, directly affecting riparian communities, aquatic biodiversity and surrounding ecosystems.

Causes of rising water levels

The rising waters of Lake Tanganyika are mainly attributed to a combination of climatic and human factors.

- **Global climate change**: Climate change is one of the main drivers of rising water levels in the great African lakes, including Lake Tanganyika. Rising global temperatures lead to an increase in water temperature, which in turn increases evaporation. However, in the case of Lake Tanganyika, increased precipitation due to climate change is outstripping that of other lakes. evaporation rates, contributing to rising water levels. Climatic variations have also disrupted traditional precipitation patterns, resulting in episodes of heavy rainfall that increase the flow of rivers feeding the lake.

- **Human activities**: The impact of human activities is also significant. Deforestation in the lake's watershed, due to agriculture, logging and urbanization, reduces the land's capacity to absorb precipitation, thus increasing runoff into the lake. In addition, the construction of infrastructure such as dams and roads alters natural water regimes, exacerbating the problem.

- **Natural variability**: The natural variability of lake levels, influenced by climatic cycles such as El Niño and La Niña, also plays a role. These phenomena can lead to periods of excessive rainfall or drought, directly affecting the level of Lake Tanganyika.

Environmental consequences

The consequences of the rising waters of Lake Tanganyika are far-reaching and varied, affecting both aquatic and terrestrial ecosystems.

- **Habitat loss**: Rising water levels submerge coastal habitats, resulting in the loss of crucial breeding and feeding areas for many species. Areas of aquatic vegetation, essential for fish and birds, are particularly hard hit.

- **Ecosystem changes**: Lake Tanganyika's ecosystems are particularly sensitive to changes in temperature and water level. Rising water levels can disrupt the lake's stratified zones, altering the habitats of endemic species and introducing invasive species that can destabilize local ecosystems.

- **Soil erosion and pollution**: Soil erosion accelerates as runoff increases, carrying sediment into the lake that can smother aquatic habitats. In addition, runoff can carry agricultural and industrial pollutants, degrading water quality and affecting the health of aquatic ecosystems.

Socio-economic impact

Human communities around Lake Tanganyika depend heavily on its resources for their livelihoods. Rising water levels pose major challenges for these populations.

- **Population displacement**: Floods submerge farmland and homes, forcing communities to move. These displacements have serious social and economic consequences, leading to the loss of livelihoods and assets, as well as heightened social tensions.

- **Food security**: Fishing is a vital activity for local populations. Rising sea levels disrupt fish breeding grounds, reducing catches and threatening the food security of fishing-dependent communities.

- **Infrastructure and economy**: Coastal infrastructures such as roads, ports and tourist facilities are damaged by erosion and flooding. Repair and reconstruction costs weigh heavily on local and national economies.

Book objectives

The aim of this book is to provide a comprehensive and detailed analysis of the causes and consequences of the rising waters of Lake Tanganyika. By combining scientific, social and economic perspectives, we hope not only to raise awareness of this critical issue, but also to propose practical solutions to mitigate its effects.

Objectives

The purpose of this book is threefold:

- **Raising awareness**: Inform a wide audience about the challenges of rising water levels in Lake Tanganyika, highlighting the complex interactions between climate change, human activities and natural variability.

- **Document**: Provide detailed, scientific documentation on the environmental and socio-economic effects of rising sea levels, based on current research and local case studies.

- **Proposing solutions**: Present adaptation and mitigation strategies that can be implemented by local communities, governments and international organizations to effectively manage rising waters.

Chapter 2: History and geography of Lake Tanganyika

2.0. Introduction

Lake Tanganyika, located in the Great Lakes region of Africa, is not only one of the largest freshwater lakes in the world, but also one of the oldest and deepest. Its formation is closely linked to the geological processes that have shaped the region over millions of years. This chapter explores the origin and formation of Lake Tanganyika, focusing on geology and plate tectonics, as well as the lake's geological history.

2.1 **Origin and formation of the lake**

Geology and plate tectonics

Lake Tanganyika lies in East Africa's Rift Valley, one of the most active tectonic zones on the planet. This region is characterized by deep faults and fractures in the earth's crust, resulting from the spreading of the African tectonic plates.

1. **Rift Valley and plate tectonics :**

The East African Rift Valley is a major geological structure stretching over 3,000 kilometers from north to south, from the Red Sea to Mozambique. It is formed by the divergence of tectonic plates, principally the African Plate, which divides into two sub-plates: the Nubian Plate to the west and the Somali Plate to the east. Lake Tanganyika lies in the western branch of this Rift Valley, known as the Albertine Rift.

2. **Formation of Lake Tanganyika:**

Lake Tanganyika was formed around 12 million years ago, during the Miocene phase of geological history. At that time, intense tectonic activity led to the subsidence of the earth's crust, creating a deep basin that would become Lake Tanganyika. The continuous divergence of tectonic plates progressively widened and deepened this basin, allowing the accumulation of water over millions of years.

3. **Geology of the Lake Tanganyika Basin :**

The Lake Tanganyika basin is surrounded by steep mountains and high plateaus, the result of tectonic compression and tension. The surrounding rocks are mainly composed of granite, gneiss and schist, which are ancient metamorphic formations. These rocks also contain sedimentary deposits that have accumulated over time, providing valuable archives of the region's climatic and ecological history.

Geological history of the lake

The geological history of Lake Tanganyika is complex and fascinating, reflecting the tectonic, climatic and biological changes that have shaped the region over millions of years.

1. Initial training :

Lake Tanganyika began to form some^ to 12 million years ago, during the Miocene. Plate tectonics created a deep, elongated basin, which gradually began to fill with water. The early phases of the lake's formation were marked by intense volcanic and tectonic activity, creating a dynamic and evolving geological landscape.

2. evolution of the lake basin :

Over the following millions of years, the Lake Tanganyika basin continued to deepen and widen. Episodes of volcanism and sedimentation added layers of rock and sediment to the basin, recording climatic and ecological changes in the region. The stratigraphy of the lake's sediments reveals periods of climatic variation, including phases of drought and wetter periods.

3. Climate change :

The geological history of Lake Tanganyika is also marked by significant climatic changes. During glacial periods, the lake level dropped considerably, sometimes almost disappearing altogether. These water level fluctuations have had a profound impact on the lake's biodiversity and ecosystems, encouraging species adaptation and diversification.

4. Fossil flora and fauna :

The sediments of Lake Tanganyika contain fossils of ancient flora and fauna, providing valuable clues to the biological history of the region. These fossils reveal a rich diversity of fish, molluscs and aquatic plants, some dating back millions of years. Studying these fossils helps us to understand the evolution of the lake's species and ecosystems over time.

5. Tectonic separation and biological isolation :

Ongoing tectonic activity in the Albertine Rift region has led to the separation and isolation of Lake Tanganyika's biological populations. This separation has favored the evolution and speciation of species, creating a unique and endemic biodiversity. Cichlid fish, for example, are renowned for their diversity and specific adaptation to the lake's different habitats.

Modern geology and lake dynamics

Today, Lake Tanganyika continues to be influenced by tectonic and geological processes. The Rift Valley remains an area of seismic and volcanic activity, with occasional earthquakes and volcanic eruptions in the region. These geological activities affect water levels, water quality and lake habitats.

1. **Seismicity and volcanoes :**

The region around Lake Tanganyika is subject to frequent seismic activity, due to ongoing tectonic movements. Earthquakes can cause underwater landslides, disturbing lake-bottom sediments and affecting water clarity. In addition, the presence of active volcanoes in the region adds a further dimension to the lake's geological dynamics.

2. **Hydrology and water circulation :**

Lake Tanganyika is also characterized by complex water circulation, influenced by temperature variations, rainfall and river inflows. The stratification of the lake's waters creates distinct zones of temperature and salinity, affecting the distribution of nutrients and aquatic organisms.

3. **Human impact and conservation :**

Human interaction with Lake Tanganyika has also left a geological and ecological imprint. Activities such as fishing, agriculture and urbanization have altered the lake's ecosystems and habitats. Conservation efforts aim to protect this unique biodiversity and mitigate the negative impacts of human activities.

Conclusion

Lake Tanganyika is a fascinating example of how geological and tectonic processes can shape a complex and dynamic ecosystem. Its formation and evolution over millions of years tell a story of change and adaptation, influenced by powerful natural forces. Understanding the geological history of Lake Tanganyika is essential to appreciate its unique biodiversity and the conservation challenges it faces today.

By exploring in depth the origin and formation of Lake Tanganyika, we can better understand the geological mechanisms that created this extraordinary lake and the natural processes that continue to shape it. This understanding is essential to protect and preserve this precious ecosystem for future generations.

2.2 Geographic and geological evolution of Lake Tanganyika

Lake Tanganyika, with its impressive dimensions and complex geological history, offers a fascinating insight into the evolution of African landscapes over millions of years. This section will explore in detail the geographical and geological changes that have shaped Lake Tanganyika, based on geological data, historical maps and current research.

2.2.1 Geographical changes over time

Lake Tanganyika is located in a geologically dynamic region, the East African Rift Valley, where tectonic activity has played a major role in its formation and evolution. This sub-section will examine the main geographical changes that have taken place over geological and historical periods.

- Geological formation :

Lake Tanganyika is one of the oldest and deepest lakes in the world, formed some 9 to 12 million years ago during the Miocene. Its origin can be traced back to intense tectonic activity in the region, characterized by the collapse of the Earth's crust along the East African Rift.

- Tectonic movements :

Movements of the African and Somali tectonic plates continue to influence the region, causing deformations and variations in the surrounding landscape. These movements have contributed to the formation of faults and the elevation of mountains around Lake Tanganyika.

- water level variability :

During glacial and interglacial periods, the level of Lake Tanganyika has fluctuated significantly in response to global climate change. These variations have left their mark on lake sediments and influenced the lake's biological diversity.

2.2.2 Historical and present-day trends

The study of historical and current maps of Lake Tanganyika offers valuable insight into its geographical evolution and the environmental changes that have affected the region over time. This sub-section will examine the use of maps to understand the evolution of Lake Tanganyika.

- Historical mapping :

The first maps drawn by European explorers in the 19th century documented the contours and size of Lake Tanganyika for the first time. These maps also reflect the limited knowledge of the region's geology and hydrology at the time.

- Modern mapping techniques :

Technological advances in cartography, such as satellite imagery and geographic information systems (GIS), now enable detailed and accurate mapping of Lake Tanganyika and its surroundings. These tools are essential for monitoring long-term environmental changes.

- Geospatial analysis :

The use of geospatial analysis enables historical data to be compared and overlaid with contemporary data, providing a dynamic view of the evolution of the lake landscape and its impact on human communities and biodiversity. By exploring these aspects of Lake Tanganyika's geographical and geological evolution in depth, we can better understand not only its complex natural history, but also the contemporary environmental challenges it faces. This section highlights the importance of geological and cartographic analysis for the sustainable management of lake resources and biodiversity conservation.

2.3 Lake Tanganyika's unique biodiversity

Lake Tanganyika is famous for harboring exceptional biodiversity, with an impressive number of endemic species adapted to its unique habitats. This section will explore in detail the endemic species, the specific ecosystems that support them, and the crucial importance of conserving this biodiversity for local ecosystems and the global community.

2.3.1   Endemic species and unique ecosystems

Lake Tanganyika is one of the world's richest freshwater ecosystems, home to many endemic species found nowhere else on the planet. This sub-section will explore the main taxonomic groups and specific ecosystems that characterize the lake's unique biodiversity.

- Endemic fish :

Lake Tanganyika is famous for its diversity of cichlids, with over 250 described species. These fish, known for their adaptability and rapid evolutionary radiation, occupy a multitude of ecological niches in the lake's deep and shallow waters.

- Other taxonomic groups :

In addition to cichlids, Lake Tanganyika is home to a variety of other endemic taxonomic groups, including aquatic invertebrates such as shrimps and snails, as well as reptiles and birds that depend on the lake's resources for their survival.

- Distinct ecosystems :

Lake Tanganyika's ecosystems vary according to depth, water turbidity and the geological composition of the lake bed. Shallow littoral zones are rich in aquatic vegetation and macroinvertebrates, while deeper waters are home to species adapted to conditions of high hydrostatic pressure and variable oxygen levels.

2.3.2   The importance of biodiversity conservation

Conservation of Lake Tanganyika's biodiversity is crucial not only to preserve unique endemic species, but also to maintain the ecological integrity of local ecosystems and sustain the livelihoods of lake-dependent human communities. This sub-section will examine the reasons why biodiversity protection is essential.

- Ecological stability :

The genetic diversity of endemic species contributes to the resilience of ecosystems to environmental disturbances such as climate change and pollution. Preserving this diversity helps to maintain the ecological balance of Lake Tanganyika and its surroundings.

- Ecosystem services :

Lake Tanganyika's ecosystems provide a multitude of ecosystem services, including regulating nutrient cycles, purifying water and providing food and economic resources for local communities. Loss of biodiversity could jeopardize these essential services.

- Global conservation :

As an IUNESCO World Heritage site and biodiversity hotspot, the conservation of Lake Tanganyika is of global importance. Preserving its endemic species contributes to global biodiversity conservation and the promotion of sustainable development on a regional and international scale. By exploring in depth these aspects of Lake Tanganyika's unique biodiversity and its importance for global conservation, we can better understand the challenges and opportunities associated with protecting this precious ecosystem.

Chapter 3: Causes of rising water levels

3.1 Introduction

Scientific observations have clearly established that the level of Lake Tanganyika is rising significantly and continuously. This worrying trend is having a direct impact on lakeside communities, their economic activities and the lake's unique environment.

Understanding the causes of rising sea levels is essential if we are to deal with them effectively and sustainably. Research carried out in recent years has identified several main factors behind this phenomenon:

1. Climate change: Rising average temperatures and changing rainfall patterns in the Lake Tanganyika basin region have led to increased freshwater inflows, fuelling a rise in lake levels.
2. Deforestation: Destruction of the forests around the lake has reduced the water-holding capacity of the soils, increasing runoff and sediment load towards the lake.
3. Human activities: expansion of agricultural areas, urbanization and industrialization in the watershed have disrupted natural hydrological balances, contributing to lake flooding.
4. Regional tectonics: Slow but continuous movements of tectonic plates beneath the lake gradually modify bathymetry and bottom topography, influencing water levels.

This chapter analyses each of these factors in detail, drawing on the latest scientific data available. It highlights the complex interplay between these different causes, underlining the importance of a holistic approach to fully understand this phenomenon.

3.1 Global climate change

Global climate change, exacerbated by human activities such as greenhouse gas emissions, is widely recognized as one of the main factors contributing to the rising water levels of Lake Tanganyika. This chapter will explore in depth the impact of global warming on the lake's water levels, using scientific data and studies to support the observed trends.

Impact of global warming on water levels

Lake Tanganyika, located in a tropical region vulnerable to climatic variations, is experiencing significant changes in its water levels in response to global warming. This sub-section will examine the mechanisms by which climate change directly affects the lake's water levels and the impacts on ecosystems and local communities.

- Melting glaciers and increased precipitation:

Global warming is leading to accelerated melting of glaciers and ice caps, increasing precipitation in the Lake Tanganyika catchment area. This translates into an increased inflow of water into the lake, directly influencing its levels and hydrological dynamics.

- variations in surface temperature :

Higher surface temperatures increase evaporation from Lake Tanganyika's surface waters. This increased evaporation can adversely affect the lake's water balance, reducing water levels and altering aquatic habitats for species sensitive to these changes.

- Changes in precipitation patterns :

Climate models indicate changes in rainfall patterns around Lake Tanganyika, with a trend towards more intense rainfall events but also prolonged periods of drought. These changes exacerbate the lake's natural water level fluctuations, affecting the stability of lake ecosystems and the livelihoods of riparian populations.

Scientific data and studies on climate trends

To fully understand the impact of climate change on Lake Tanganyika, this section will examine the data and conclusions of recent scientific studies on regional and global climate trends.

- historical observations :

Historical data show a progressive increase in mean annual temperatures in the Lake Tanganyika catchment area. These trends are corroborated by long-term meteorological records and paleoclimatic data from lake sediment cores.

- Climate modeling :

Global climate models predict future scenarios in which temperatures will continue to rise, accompanied by changes in rainfall patterns. These models help to project the potential impacts of climate change on Lake Tanganyika's water resources and the communities that depend on them.

- Effects on aquatic ecosystems :

Studies of Lake Tanganyika's aquatic ecosystems indicate increased sensitivity to climate variations, with potential implications for the biodiversity of endemic species and the productivity of fisheries. Rising water temperatures and changes in rainfall patterns can disrupt the reproductive cycles and habitats of aquatic species.

By exploring these aspects of global climate change and its impact on Lake Tanganyika's water levels, we can better understand the environmental challenges facing this region and the actions needed to mitigate these effects and promote sustainable management of lake resources.

3.2 Impact of human activities

Human activities around Lake Tanganyika play a crucial role in the ecological and hydrological balance of this unique region. This chapter will explore in detail the various impacts of human activities on the lake's rising waters, focusing on deforestation and land use, agriculture and urbanization, as well as pollution and water resource management.

3.2.1 Deforestation and land use

Deforestation and the modification of surrounding land have a significant impact on the hydrology and ecology of the Lake Tanganyika watershed. This sub-section will examine the mechanisms by which these activities affect the lake's water levels and potential solutions to mitigate these effects.

- **Deforestation and soil erosion:**

  Deforestation along the shores of Lake Tanganyika increases soil erosion and rainwater runoff. This increases the amount of sediment transported into the lake, affecting water quality and the habitat of aquatic species.

- **Modification of terrestrial habitats :**

  The expansion of agriculture and urban areas is altering the natural habitats around Lake Tanganyika, fragmenting terrestrial ecosystems and reducing the natural buffers that absorb rainwater. This can intensify flooding during the rainy season and reduce water flows during the dry season.

- **Consequences for biodiversity :**

  Habitat loss due to deforestation and unsustainable land use is threatening terrestrial and aquatic biodiversity around Lake Tanganyika. Some endemic species depend on forest ecosystems for their survival, while others are directly affected by deteriorating water quality.

3.2.2   Agriculture and urbanization

Intensive agriculture and urban expansion around Lake Tanganyika also contribute to the environmental pressures that influence the lake's water levels. This sub-section will explore the specific effects of these activities on local hydrology and the human communities that depend on it.

- Intensive farming :

Excessive use of fertilizers and pesticides on farms can lead to pollution of surface and groundwater, affecting water quality in Lake Tanganyika and reducing the availability of water resources for other uses.

- Growing urbanization :

The expansion of urban areas around Lake Tanganyika is leading to increased demand for drinking water and sanitation services. This additional pressure on water resources can compromise the availability of freshwater for aquatic ecosystems and local communities.

- Watershed management :

Sustainable watershed management is essential to mitigate the effects of agriculture and urbanization on water levels in Lake Tanganyika. Sustainable agricultural practices and planned urbanization strategies can minimize negative impacts on water resources and support local biodiversity.

3.2.3　Pollution and water resource management

Pollution, whether from agricultural, urban or industrial activities, is a growing threat to the quality of Lake Tanganyika's water and to the health of aquatic and human ecosystems. This sub-section
will examine the sources of pollution and the challenges associated with sustainable water resource management.

- Nutrient pollution :

Excess nutrients from agricultural and urban wastewater can cause algal blooms and dead zones in Lake Tanganyika. This affects water quality and disrupts natural ecological cycles, compromising biodiversity and fisheries productivity.

- Chemical pollution :

Unregulated industrial discharges and mining practices can introduce chemical contaminants into Lake Tanganyika, endangering human and environmental health. Monitoring and regulating polluting activities are essential to prevent harmful impacts on lake ecosystems.

- Integrated water resource management :

An integrated approach to water resource management, involving governments, local communities and industry players, is needed to ensure the sustainability of Lake Tanganyika's water resources. This includes implementing strict environmental standards, promoting clean technologies and raising public awareness of the importance of conserving water resources.

By exploring in depth these aspects of the impact of human activities on the rising waters of Lake Tanganyika, we can better understand the complex environmental challenges facing this region and identify the strategies needed to promote sustainable management and effective conservation of natural resources.

3.3 Natural water level variability

As one of the largest freshwater lakes and the second deepest lake in the world, Lake Tanganyika is subject to complex natural water-level variability. This chapter examines in detail the natural cycles that influence water level fluctuations, as well as the impact of rainfall and drought on these hydrological dynamics.

3.3.1 Natural cycles and interannual variability

The natural variability of water levels in Lake Tanganyika is influenced by a combination of climatic and hydrological factors, including seasonal cycles and longer-term variations. This subsection explores the main cycles and interannual variability observed in the lake's water levels.

- Seasonal cycles :

Lake Tanganyika experiences seasonal fluctuations in water levels, generally marked by periods of high water during the rainy season and low water during the dry season. These variations are mainly influenced by rainfall patterns in the lake's watershed.

- Interannual variability :

In addition to seasonal cycles, Lake Tanganyika exhibits inter-annual variability, where water levels can vary from year to year in response to global climatic conditions such as El Niño and La Niña. These climatic phenomena influence rainfall patterns in the region and have a direct impact on the lake's water resources.

- **Historical data and observations** :

Historical records and hydrological studies offer insights into the long-term variability of water levels in Lake Tanganyika. Lake sediment cores and geophysical data help to reconstruct past variations and project future trends in water levels.

3.3.2 Influence of precipitation and drought

Rainfall and drought play a crucial role in the hydrological dynamics of Lake Tanganyika, directly influencing water levels and the biodiversity of lake ecosystems. This sub-section examines the impact of variations in rainfall patterns and periods of drought on the lake.

- **Precipitation patterns** :

Rainfall patterns in the Lake Tanganyika catchment area vary considerably from year to year and season to season. Years of heavy rainfall increase water inflows into the lake, while years of drought reduce water levels and can lead to water crises for riparian communities.

- **Effects of drought** :

Prolonged periods of drought can have devastating effects on lake ecosystems and the human communities that depend on Lake Tanganyika. Reduced water flows and diminished availability of water resources can compromise biodiversity, food security and the livelihoods of local populations.

- **Climate change modeling**

Global climate models project future changes in rainfall patterns and drought occurrence in the Lake Tanganyika catchment area. These projections are essential for anticipating the potential impacts of climate change on the lake's water levels, and for formulating appropriate adaptation and mitigation strategies.

These aspects of natural water level variability in Lake Tanganyika enable us to better understand the complex hydrological dynamics that govern this unique lake system. This understanding is crucial for informing water resource management policies and promoting the resilience of ecosystems and local communities to future environmental challenges.

Conclusion

The detailed analysis carried out in this chapter has identified the main causes of the worrying rise in Lake Tanganyika's water level since the early 2010s. It is clear that this phenomenon is the result of a combination of several interdependent factors, whether climatic, hydrological or linked to human activities.

Climate change, with rising temperatures and changing precipitation patterns, plays a central role in amplifying freshwater inputs. Intensive deforestation in the watershed has also reduced the water retention capacity of the soils, aggravating runoff into the lake. At the same time, human activities such as agricultural expansion, urbanization and industrialization have disrupted the natural hydrological balance.

What's more, the slow but continuous movement of tectonic plates beneath the lake is gradually modifying its bathymetry and topography, also influencing water levels.

This multifactorial understanding of the causes of rising water levels in Lake Tanganyika is essential if we are to define and implement effective and sustainable adaptation and mitigation strategies. This is not a simple phenomenon to grasp, but a complex challenge requiring a global and integrated approach.

Chapter 4: Environmental consequences

Introduction

Lake Tanganyika, with its unique biodiversity and crucial ecological importance, has been facing a major challenge since the early 2010s, as outlined in our introduction: the continuing rise in its water level. As we saw in the previous chapter, this phenomenon is the result of a complex combination of factors, including climate change, deforestation and human activities.

Beyond the direct socio-economic impacts on riparian communities, this rising water level also has profound and worrying environmental repercussions throughout the Lake Tanganyika basin region. Understanding these consequences is essential to defining effective adaptation and protection strategies.

This chapter focuses on a detailed analysis of the main environmental impacts of rising lake levels, based on the latest scientific data available. It highlights the transformations underway in aquatic and terrestrial ecosystems, as well as the threats to biodiversity and ecosystem services crucial to local populations.

Particular attention will be paid to the cascading effects of these environmental changes, underlining the need for a holistic and integrated approach to these complex challenges. Only a thorough understanding of the impact of rising sea levels will enable us to implement sustainable solutions that are in tune with the realities on the ground.

4.1   Impact on aquatic flora and fauna

The rising waters of Lake Tanganyika are having a significant impact on aquatic ecosystems, affecting the biodiversity and population dynamics of fish and other endemic species. This chapter explores in detail the environmental consequences of these changes, focusing on the specific impacts on aquatic fauna and flora.

4.1.1   Changes in fish populations and other aquatic species

Lake Tanganyika is renowned for its unique biodiversity of fish, including many endemic species adapted to its habitats.

populations. Rising sea levels directly affect these populations, influencing their habitat and breeding strategies.

- Species distribution :

Changes in water levels alter the geographical distribution of fish species in Lake Tanganyika. Shoreline habitats may be submerged, forcing fish to move to deeper areas or seek out new, suitable habitats.

- Migration and reproduction :

Variations in water levels impact the migration and reproduction processes of fish in Lake Tanganyika. Species that depend on littoral zones for reproduction can be disrupted, compromising the success of their life cycles and threatening the survival of populations.

- Competition and predators :

Environmental changes induced by rising water levels can intensify competition between species and increase the pressure exerted by predators on fish populations. This may disrupt the pre-existing ecological balance in Lake Tanganyika.

4.1.2 Loss of natural habitats

The rising waters of Lake Tanganyika are causing the submergence of coastal areas and the loss of critical natural habitats for many aquatic species. This loss of habitat affects not only biodiversity but also the resilience of lake ecosystems to environmental pressures.

- Bank erosion

Rising water levels can accelerate shoreline erosion on Lake Tanganyika, reducing the availability of shoreline habitats for aquatic plants and animal species. This erosion compromises shoreline stability and disrupts natural ecological cycles.

- Loss of spawning areas :

Fish species in Lake Tanganyika often depend on specific areas for spawning and larval development. Submerging these spawning areas can reduce reproduction rates and increase the vulnerability of fish populations to anthropogenic and environmental pressures.

- Impact on biodiversity

The disappearance of natural habitats is threatening the genetic and ecological diversity of Lake Tanganyika. Endemic and specialized species may lose their unique ecological niches, increasing their risk of local or regional extinction.

4.2 Changes in coastal ecosystems

Lake Tanganyika's coastal ecosystems are critical areas that harbor unique biodiversity and provide essential ecosystem services to local communities. This chapter looks in detail at the impacts of environmental change, particularly rising sea levels, on these sensitive coastal ecosystems.

4.2.1 Coastal erosion and land loss

Coastal erosion is a worrying phenomenon around Lake Tanganyika, exacerbated by rising sea levels and other anthropogenic factors. This sub-section explores the causes and consequences of coastal erosion, as well as the challenges associated with the sustainable management of these vulnerable areas.

- Causes of coastal erosion :

Rising sea levels due to climate change are contributing directly to coastal erosion on Lake Tanganyika. Stronger waves and rising water levels compromise shoreline stability, increasing the rate of erosion.

- Effects on local communities :

Coastal erosion is threatening the infrastructure, housing and farmland of communities along the shores of Lake Tanganyika. The loss of fertile land and reduced access to fresh water exacerbate the socio-economic pressures on these vulnerable populations.

- Mitigation strategies :

Techniques such as the construction of coastal protection structures, the planting of stabilizing plant species and integrated watershed management are essential to mitigate the effects of coastal erosion and restore the resilience of coastal ecosystems.

4.2.2 Modification of wetlands and coastal habitats

Wetlands play a crucial role in hydrological regulation, water filtration and the preservation of biodiversity around Lake Tanganyika. This sub-section examines how rising water levels and other anthropogenic pressures are altering these essential habitats.

- Wetland reduction :

Rising sea levels are submerging and reducing the wetlands around Lake Tanganyika. These valuable habitats are essential for many aquatic and terrestrial species, and their loss compromises ecological stability and the ability of ecosystems to provide ecosystem services.

- Alteration of coastal habitats :

Changes in water levels influence the structure and composition of Lake Tanganyika's coastal habitats. Plant species adapted to the specific conditions of coastal areas may be disturbed, affecting their ability to provide habitats and breeding grounds for local fauna.

- Conservation of coastal ecosystems

Protecting wetlands and coastal habitats is crucial to maintaining biodiversity and ensuring the resilience of Lake Tanganyika's ecosystems. Integrated management strategies, such as the establishment of nature reserves and the restoration of degraded ecosystems, are needed to preserve these sensitive areas.

4.3 Risks for endemic species

Lake Tanganyika is a treasure trove of biodiversity, home to a multitude of unique endemic species. However, this biological wealth is currently threatened by a variety of anthropogenic and environmental factors. Endemic species, which have evolved over millions of years in this isolated lake, are particularly vulnerable to these factors.

threats. This section explores in detail the specific dangers they face and the efforts being made to protect them.

Threats specific to endemic species
1. Climate change

Climate change is a major threat to the endemic species of Lake Tanganyika. Rising water temperatures are affecting aquatic habitats and altering the living conditions of sensitive species. For example, studies have shown that rising water temperatures can disrupt the reproduction of endemic cichlid fish, leading to a decline in their populations (Smith et al., 2021, p. 245-260).

2. Water pollution

Water pollution due to human activities is another threat factor. Agricultural chemical discharges, industrial effluents and domestic wastewater contaminate the lake, altering water quality and affecting the health of endemic species. Pollutants can cause deformities, disease and even death in fish and other aquatic organisms (Johnson, 2020, p. 175-190).

3. Introduction of non-native species

The introduction of non-native species into Lake Tanganyika has devastating impacts on endemic species. Invasive species can compete for resources, predate on local species or introduce new diseases. A notable example is the introduction of the Nile perch, which has already caused havoc in other African lakes (Brown, 2018, p.j0-105).

4. Overfishing

Overfishing is a direct threat to endemic fish populations. Unsustainable fishing practices, such as the use of fine-mesh nets, capture juvenile fish before they have a chance to reproduce. This pressure on fish populations leads to a decline in species, with some even threatened with extinction (Green, 2019, p. 80-95).

5. Habitat destruction

Habitat destruction, particularly through deforestation and coastal development, contributes to biodiversity loss. Fish spawning grounds and natural habitats are disrupted, reducing the chances of survival of endemic species. Sediments from soil erosion can also smother aquatic habitats and reduce water clarity, affecting the photosynthesis of aquatic plants and the overall health of ecosystems (Taylor, 2023, p. 210-225).

Ongoing conservation efforts

1. Habitat restoration projects

Various habitat restoration projects have been launched to protect and rehabilitate Lake Tanganyika's aquatic ecosystems. These initiatives include the replanting of riparian vegetation to reduce soil erosion and improve water quality. Marine Protected Areas (MPAs) are also being created to provide safe havens for species to reproduce (Doe, 2021, p. 135-150).

2. Monitoring and research programs

Continuous monitoring and scientific research are essential to understand population dynamics and the impact of threats on endemic species. Monitoring programs involve tagging fish, collecting water quality data and carrying out detailed ecological studies. This information is used to formulate adaptive management strategies for species conservation (Johnson, 2020, p. 200215).

3. Community awareness initiatives

Raising awareness among local communities of the importance of Lake Tanganyika's biodiversity and of sustainable conservation practices is crucial. Educational workshops, awareness campaigns and community engagement programs are organized to encourage the active participation of local people in the conservation of endemic species. By involving local communities, these initiatives aim to promote sustainable fishing practices and reduce negative environmental impacts (Smith et al., 2021, p. 275-290).

4. Environmental policies and regulations
The governments of the countries bordering Lake Tanganyika have put in place policies and regulations to protect endemic species. These include strict regulation of fishing, protection of sensitive areas and pollution control. Regional and international cooperation also plays a key role in implementing these policies and building capacity for natural resource conservation (Brown, 2018, p. 150-165).

5. International partnerships

Conservation efforts also benefit from partnerships with international organizations, NGOs and research institutions. These collaborations provide financial resources, technical support and expertise for the implementation of large-scale conservation projects. Initiatives such as the United Nations Environment PROGRAMME (UNEP) and the Convention on Biological Diversity (CBD) play a central role in protecting Lake Tanganyika's

biodiversity.
(Green, 2019, p. 180-195).

Conclusion

Preserving the endemic species of Lake Tanganyika requires an integrated and collaborative approach. Specific threats such as climate change, water pollution, introduction of non-native species, overfishing and habitat destruction need to be addressed holistically. Ongoing conservation efforts, including habitat restoration projects, monitoring and research programs, community awareness initiatives, environmental policies and international partnerships, play a crucial role in protecting this unique biodiversity. The cooperation and commitment of all stakeholders is essential to ensure the long-term survival of Lake Tanganyika's endemic species.

Furthermore, the in-depth analysis carried out in this chapter has highlighted the profound and interconnected environmental impacts caused by the continuing rise in Lake Tanganyika's water level. The aquatic and terrestrial ecosystems of this Central African region are facing major transformations, seriously threatening the unique biodiversity and ecosystem services essential to local populations.

The gradual submergence of coastal areas, the erosion of riverbanks, the degradation of natural habitats and the disruption of hydrological cycles have cascading repercussions throughout the watershed. The survival of numerous endemic species of fish, birds, mammals and plants is now compromised, jeopardizing the fragile balance of these ecosystems.

In addition to the loss of biodiversity, these environmental transformations are seriously affecting the livelihoods and food security of lakeside communities, who are closely dependent on the natural resources of the lake and its shores. The situation is becoming critical and requires urgent, concerted action to reverse these worrying trends.

The next steps must focus on implementing integrated water resource and ecosystem management strategies, closely involving local stakeholders. Efforts to restore, conserve and protect sensitive areas must be undertaken as a matter of urgency, drawing on best scientific practice and traditional knowledge.

Only a holistic, watershed-wide approach will enable us to meet this major environmental challenge and preserve Lake Tanganyika's unique ecological wealth for future generations. Time is running out, and action must match the urgency of the situation.

Chapter 5: Socio-economic impacts

Introduction

Lake Tanganyika, with its rising waters, poses significant challenges to riparian communities. This chapter explores in detail the impact of these changes on local populations, including forced displacement, changes in lifestyles and economic activities. Recent data on flooding of homes and other infrastructure illustrate the scale of these impacts.

5.1 Impact on riverside communities

Communities along the shores of Lake Tanganyika are facing increasing socio-economic challenges as a result of rising water levels. Flooding caused by rising water levels directly affects homes, infrastructure and the livelihoods of local populations. These repercussions are particularly severe in areas where populations rely heavily on the lake's resources for their daily survival.

Population displacement

One of the most dramatic consequences of the rising waters of Lake Tanganyika is the displacement of people living along its shores. Regular flooding and shoreline erosion are forcing many families to abandon their homes and land.

Displacement scale

According to a recent study by the University of Dar es Salaam, some 15,000 people have been displaced in the coastal regions of Burundi, Tanzania, the Democratic Republic of Congo (DRC) and Zambia over the past ten years. This number is steadily increasing due to the growing impacts of climate change and inadequate land management practices.

Testimonies from affected populations

Testimonies gathered in the Kigoma region of Tanzania illustrate the seriousness of the situation. Mariam, a 35-year-old resident, recounts: "Our house was submerged by the waters of the lake last year. We lost everything we had. Now we live with relatives in a higher area, but life is very difficult."

Travel problems

Displaced families face numerous problems, including loss of livelihoods, limited access to basic services such as drinking water, education and healthcare, and potential conflicts with host communities over limited resources. Displacement also leads to social tensions and psychological disruption, particularly for children and the elderly.

Changes in lifestyles and economic activities

The rising waters of Lake Tanganyika are profoundly altering the lifestyles and economic activities of the people living along its shores. These changes mainly affect fishing, agriculture and commercial activities linked to the lake.

Impact on fishing

Fishing, a vital economic activity for communities around Lake Tanganyika, is being severely affected by rising water levels. Local fishermen are witnessing a decline in fish stocks and changes in the behavior of aquatic species.

-Reduced fish stocks: Flooding disrupts natural fish habitats, leading to reduced populations of some species. Fishermen are reporting ever-lower catches, threatening their main source of income and food. A study by the FAO (Food and Agriculture Organization of the United Nations) indicates that fish catches in some areas of the lake have fallen by 20% over the past five years.

- Changes in fish behaviour: Variations in water levels alter fish migration patterns and spawning grounds. This makes fishing more unpredictable and forces anglers to constantly adapt, often using less efficient and more costly techniques.

Impact on agriculture

Agriculture, another economic mainstay of lakeside communities, is also suffering the consequences of rising water levels. Farmland close to the lake's shores is often flooded, reducing the area available for crops.

- Loss of farmland: Many parcels of arable land are submerged, leading to a drop in agricultural production. Farmers often have to move to less fertile land, reducing yields and increasing food insecurity. According to a World Bank report, around 10% of agricultural land in riparian zones has been lost to flooding over the past ten years.

- Soil degradation: Flooding also leads to soil degradation through erosion and salinization, making the land less productive. This makes it even harder for farmers to maintain sufficient production levels.

Impact on infrastructure and housing

Essential infrastructure such as roads, bridges and sanitation facilities also suffer considerable damage from flooding. The homes of riverside populations are particularly vulnerable.

- **House flooding**: According to data from the National Institute of Statistics of the DEMOCRATIC REPUBLIC OF THE CONGO, over 5,000 houses have been flooded or damaged in the Kivu region over the past three years. The families affected lose not only their homes, but also their personal possessions, exacerbating their economic precariousness.

- **Damage to infrastructure**: Flood-damaged roads and bridges complicate transport and access to markets, schools and health centers. This hampers the mobility of people and goods, increasing economic costs and reducing communities' quality of life.

Community adaptation and resilience

Faced with these challenges, communities living along the shores of Lake Tanganyika are developing adaptation and resilience strategies to cope with the impacts of rising water levels.

- **Local coping strategies**: Local populations implement measures such as building dykes and terraces to protect farmland, planting vegetation to stabilize soils and diversifying sources of income to reduce dependence on a single economic activity.

- **Role of NGOs and local institutions**: Many non-governmental organizations (NGOs) and local institutions work with communities to develop training and technical assistance programs. For example, the "Resilient Livelihoods" project run by the NGO World Vision in Burundi helps farmers to adopt sustainable agricultural practices and improve their access to markets.

- **Governmental and international support**: The governments of the countries bordering Lake Tanganyika, with the support of the international community, are implementing policies and programs to strengthen community resilience. This includes setting up flood early warning systems, building climate-proof infrastructure and promoting the sustainable management of natural resources.

Conclusion

The rising waters of Lake Tanganyika are having a profound socio-economic impact on riverside communities, leading to forced displacement, changes in lifestyles and economic disruption. Data on house flooding and testimonies from affected populations highlight the scale of these challenges. Despite this, local communities are demonstrating remarkable resilience by developing innovative adaptation strategies, supported by local, governmental and international initiatives. It is essential to continue strengthening these efforts to ensure the sustainability and prosperity of local populations in the face of the impacts of climate change.

5.3 Adaptation and resilience of local populations

Faced with the challenges posed by the rising waters of Lake Tanganyika, riparian communities must develop adaptation and resilience strategies to survive and prosper. This chapter looks in detail at local adaptation initiatives, and the crucial role played by local NGOS and institutions in supporting these efforts.

Local adaptation strategies

Communities along Lake Tanganyika are adopting a variety of strategies to adapt to the impacts of rising water levels. These strategies are often based on traditional knowledge and proven practices, but also include modern innovations to meet new challenges.

Water management and land protection

Efficient water management and land protection against erosion and flooding are crucial aspects of local adaptation strategies.

- **Building dikes and terraces**: Communities build dikes and terraces to protect farmland from flooding. These structures help control the flow of water and prevent soil erosion. For example, in the Bujumbura region of Burundi, farmers use sandbags and stones to create barriers along the banks of the lake.

- **Reforestation and planting vegetation**: Reforesting coastal areas with native tree species and planting water-resistant vegetation help stabilize soils and reduce erosion. Trees also play an important role in regulating the water cycle and protecting natural habitats. Tree-planting programs in villages around Kigoma in Tanzania are showing positive results in terms of reducing erosion and improving ecological resilience.

Livelihood diversification

To reduce their dependence on a single source of income, local communities are diversifying their economic activities. This creates alternative sources of income and strengthens economic resilience in the face of. environmental impacts.

- **Sustainable agriculture and aquaculture**: In addition to traditional fishing, communities are investing in aquaculture and adopting sustainable farming practices. Aquaculture offers a viable alternative by raising fish in controlled environments, reducing pressure on wild fish stocks. Farmers are also adopting techniques such as agroforestry, crop rotation and the use of plant varieties resistant to variable climatic conditions.

- **Non-agricultural economic activities**: Communities are turning to non-agricultural economic activities, such as handicrafts, retailing and tourism. For example, in the Mpulungu region of Zambia, women make baskets and handicrafts from local materials, which they sell to tourists and at local markets.

Improving infrastructure and access to services

Improving infrastructure and access to essential services are key to strengthening community resilience to climate impacts.

- **Flood-resistant infrastructure**: Communities invest in the construction and repair of flood-resistant infrastructure, such as elevated roads, robust bridges and efficient drainage systems. These improvements facilitate mobility and access to markets, schools and health centers, even during periods of heavy rain and flooding.

- **Access to education and health**: Access to education and health services is essential for building resilience. Environmental education and awareness programs are set up to inform communities about best practices in natural resource management and adaptation strategies. Local health centers play a crucial role in providing medical care and raising awareness of the health risks associated with flooding and environmental degradation.

Role of NGOs and local institutions

Local NGOs and institutions play a key role in supporting the adaptation and resilience strategies of communities living along Lake Tanganyika. They provide technical, financial and logistical assistance, while building local capacity to manage environmental impacts in a sustainable way.

Technical assistance and capacity building
Local NGOs and institutions provide technical expertise and training to help communities adopt effective adaptation practices.

- **Training and awareness-raising**: Training programs are organized to raise community awareness of the impacts of climate change and best practices in resource management. For example, the NGO CARE International organizes training workshops in Tanzania on agroforestry, water management and flood-resistant construction techniques.

- **Skills development**: Skills development programs enable local populations to acquire new skills and knowledge to diversify their livelihoods. The NGO World Vision, for example, offers training in aquaculture, entrepreneurship and financial management to communities bordering Lake Tanganyika.

Financial assistance and access to resources
NGOs and local institutions also provide financial support and facilitate access to the resources needed to implement adaptation strategies.

- **Microcredit and grants**: Microcredit and grant programs enable communities to access financial resources to invest in adaptation projects. For example, the microcredit program of the NGO Heifer International in Burundi helps farmers finance initiatives such as the purchase of drought-resistant seeds and the construction of rainwater harvesting systems.

- **Access to technology**: NGOs facilitate access to modern technologies and equipment needed to improve community resilience. For example, Practical Action provides solar irrigation pumps and water filtration systems to communities along Lake Tanganyika.

Strengthening infrastructure and services
Local NGOS and institutions work with governments and communities to improve infrastructure and essential services.

- **Infrastructure projects**: Infrastructure projects supported by NGOs include the construction of dykes, elevated roads, drainage systems and community centers. These projects improve the resilience of local infrastructure to flooding and facilitate access to basic services. For example, the NGO International Rescue Committee (IRC) has built flood-resistant community centers in the DRC, where residents can take refuge during periods of high water.

- **Improving health and education services**: NGOs support health and education systems by providing equipment, training and resources. The NGO Médecins Sans Frontières (MSF) works in flood-affected areas to provide emergency health care and prevent outbreaks of water-borne diseases.

Advocacy and political influence
Local NGOs and institutions also play an important role in advocating sustainable development policies and programs.

- **Advocacy for climate policies**: NGOS work with local and national governments to promote climate policies that support communities' adaptation and resilience strategies. They advocate the integration of sustainable natural resource management into development plans, and the implementation of social protection programs for vulnerable populations.

- **Community mobilization**: NGOs encourage the active participation of local communities in decision-making and resource management. They organize community forums and discussion groups to gather the opinions of local residents and involve them in the planning and implementation of adaptation projects.

Collaboration and partnerships
Collaboration between NGOS, local institutions, governments and communities is essential to maximize the impact of adaptation and resilience initiatives.

- **Public-private partnerships**: Public-private partnerships (PPPs) facilitate the sharing of resources, skills and technologies between the public and private sectors. For example, a PPP between the Tanzanian government, the NGO

WWF and local companies support reforestation and watershed protection projects around Lake Tanganyika.

- **Regional cooperation**: Regional cooperation between the countries bordering Lake Tanganyika is crucial for transboundary resource management and the implementation of coordinated adaptation strategies. The Lake Tanganyika Authority (LTA) plays a key role in facilitating cooperation between Tanzania, Burundi, the DRC and Zambia for the sustainable management of the lake and its resources.

Conclusion

People living along the shores of Lake Tanganyika are demonstrating remarkable resilience in the face of the challenges posed by rising water levels, by developing innovative adaptation strategies and diversifying their livelihoods. Local NGOS and institutions are playing a crucial role in supporting these efforts, providing technical, financial and logistical assistance, strengthening

Chapter 6: Answers and solutions Introduction

The rising waters of Lake Tanganyika pose major environmental, economic and social challenges for neighboring communities and countries. Faced with these challenges, it is essential to implement effective responses and solutions, both locally and internationally. This chapter explores various resource management initiatives, adaptation and mitigation strategies, and the importance of regional and international cooperation in responding to this complex phenomenon.

Local and international resource management initiatives focus on ecosystem conservation, sustainable natural resource management and support for affected communities. These projects include efforts to improve the resilience of local populations, restore degraded habitats and promote sustainable practices. Cooperation between the countries bordering Lake Tanganyika is crucial to the implementation of these initiatives, as the lake crosses several national borders and the actions of one country can have repercussions on the whole ecosystem.

Adaptation and mitigation strategies aim to reduce the negative impacts of rising sea levels and strengthen the capacity of communities to cope with these changes. These measures include building resilient infrastructure, diversifying livelihoods, and improving disaster risk management. Emissions reduction and resource management policies also play a key role in mitigating the effects of climate change, which is contributing to the lake's rising water levels.

Finally, regional and international cooperation is essential to coordinate conservation efforts, share knowledge and resources, and ensure integrated management of the Lake Tanganyika basin. Regional programs and agreements, as well as the role of international organizations, are essential to ensure an effective and sustainable response to rising waters.

This chapter highlights the various responses and solutions implemented to cope with the rising waters of Lake Tanganyika, focusing on local and international initiatives, adaptation and mitigation strategies, and the importance of regional and international cooperation.

These concerted efforts are essential to protect Lake Tanganyika's unique ecosystem and ensure a sustainable future for the communities along its shores.

6.1 Local and international resource management initiatives

The rising waters of Lake Tanganyika represent a serious threat to riparian communities, the ecosystem and local economic activities. In response to this crisis, various local and international initiatives have been put in place to manage resources sustainably, mitigate negative impacts and build community resilience. This chapter explores these initiatives in detail, highlighting conservation projects, sustainable management and cooperation between countries bordering Lake Tanganyika.

Conservation and sustainable management projects

Conservation and sustainable management projects play a crucial role in protecting Lake Tanganyika's ecosystem and supporting local populations. These initiatives vary in scope and method, including reforestation efforts, fisheries management programs and initiatives to improve water quality.

Reforestation and watershed management projects

Deforestation around Lake Tanganyika contributes significantly to soil erosion and watershed degradation, exacerbating flooding and rising water levels. To counter these effects, several reforestation projects have been launched.

- **WWF reforestation project**: The World Wide Fund for Nature (WWF) has implemented a reforestation project in the Kigoma region of Tanzania. The project aims to reforest degraded areas with native tree species, such as Acacia and Moringa, which are resistant to local conditions and beneficial for soil stability. By 2022, over 500,000 trees have been planted, covering an area of 200 hectares.

- **LATAWAMA watershed management initiative**: THE NGO LATAWAMA (Lake Tanganyika ^vater ^4anage^rent Authority) has developed a watershed management initiative that includes reforestation activities and the promotion of sustainable agricultural practices. Working with local communities, LATAWAMA has restored over 100 hectares of eroded land around the lake by planting trees and building terraces to reduce erosion.

Fisheries management programs

Fishing is a vital economic activity for coastal communities, but it is threatened by overfishing and the environmental impacts of rising sea levels.

To ensure the sustainability of fisheries, various fisheries management programs have been set up.

- **FAO fisheries management program**: The Food and Agriculture Organization of the United Nations (FAO) has launched a fisheries management program in the Lake Tanganyika basin. This program focuses on training local fishermen in sustainable fishing practices, monitoring fish stocks and implementing regulations to control fishing. Since the program began in 2018, catches of certain fish species have shown signs of stabilizing.

- **Community fishing cooperatives**: In several regions around the lake, community fishing cooperatives have been formed to manage resources collectively and sustainably. These cooperatives set fishing quotas, monitor illegal fishing activities and work in partnership with local authorities to protect aquatic habitats. In Uvira, DRC, the "Uvira Pêche Durable" fishing cooperative has succeeded in reducing destructive fishing practices and increasing fishermen's incomes through effective community management.

Initiatives to improve water quality
The quality of Lake Tanganyika's water is essential to the health of the ecosystem and the human populations that depend on it. Several local and international initiatives aim to reduce water pollution and improve water management.

- **AFPDE "Eaux Propres" initiative**: The Association des Femmes pour la Promotion et le Développement Endogène (AFPDE) has set up the "Eaux Propres" initiative to raise community awareness of the need to protect water quality and promote environmentally-friendly waste management practices. In collaboration with local schools, AFPDE organizes lake shore clean-up campaigns and waste management workshops. Since the launch of the initiative in 2021, pollution levels from

Cooperation between countries bordering Lake Tanganyika
The sustainable management of Lake Tanganyika requires cross-border cooperation due to its location, which is shared by several countries. Cooperation between riparian countries is essential to coordinate conservation efforts, manage resources equitably and respond to common environmental challenges.

Lake Tanganyika Authority (ALT)

The Lake Tanganyika Authority (LTA) is an intergovernmental organization created to promote the sustainable management of the lake and its resources. It facilitates cooperation between Tanzania, Burundi, the DRC and Zambia.

- **Environmental cooperation protocol**: In 2008, the riparian countries signed an environmental cooperation protocol under the aegis of the ALT. This protocol provides for the implementation of common policies for biodiversity conservation, fisheries management and pollution control. It also establishes monitoring and evaluation mechanisms to measure the impact of environmental initiatives.

- **Joint monitoring program**: L`ALT coordinates a joint monitoring program to collect data on water quality, fish stocks and environmental changes in the lake. Data is shared between member countries to facilitate integrated management and inform policy decisions. In 2022, a joint assessment revealed an improvement in water quality in some parts of the lake thanks to efforts to reduce pollution.

Cross-border conservation projects
Cross-border conservation projects are implemented to protect ecosystems and promote sustainable management of natural resources.

- **Ruzizi delta conservation project**: The Ruzizi delta, located on the border between the DRC and Burundi, is an ecologically significant wetland. A cross-border conservation project, supported by ALT and funded by the United Nations Environment Programme (UNEP), aims to protect this area by rehabilitating degraded habitats, building the capacity of local managers and involving communities in conservation activities. By 2021, the project has succeeded in restoring more than

50 hectares of wetlands and increase the population of certain species of aquatic birds.

- **Cross-border reforestation initiative**: riparian countries collaborate on cross-border reforestation projects to combat soil erosion and improve watershed resilience. A notable example is the reforestation project run by Tanzania and Zambia, which aims to reforest degraded areas along their shared border. By 2022, over 300,000 trees had been planted as part of this project, covering an area of 150 hectares.

International partnerships for financing and technical support

International partnerships play a key role in funding and providing technical support for resource management initiatives around Lake Tanganyika.

- **Global Environment Facility (GEF)**: The GEF funds several conservation and sustainable management projects around Lake Tanganyika. For example, the GEF-funded Tanganyika Biodiversity project aims to protect the lake's unique biodiversity by building the capacity of local managers, supporting community-based conservation initiatives and improving resource management policies. Since its launch in 2015, the project has contributed to the creation of several protected areas and the training of hundreds of resource managers.

- **Collaboration with universities and research institutes**: Partnerships with universities and research institutes provide valuable technical support for resource management projects. For example, the University of Dar es Salaam is collaborating with ALT managers to research the impacts of climate change on Lake Tanganyika and develop science-based adaptation strategies.

Conclusion

Local and international resource management initiatives around Lake Tanganyika demonstrate the importance of cooperation and the implementation of conservation and sustainable management strategies. Reforestation projects

6.2 Adaptation and mitigation strategies

The rising waters of Lake Tanganyika, exacerbated by climate change, pose significant challenges for riparian communities and the ecosystem. To meet these challenges, it is crucial to develop adaptation strategies to minimize negative impacts and mitigation measures to reduce the underlying causes of climate change. This chapter looks in detail at adaptation measures in the face of rising sea levels, as well as emission reduction and resource management policies.

Adapting to rising water levels

Adaptation measures are designed to help communities cope with the current and future impacts of rising sea levels. They include technical interventions, adjustments to agricultural practices and community planning strategies.

Flood-resistant infrastructure
Building and renovating flood-resistant infrastructure is an essential strategy for protecting communities and their assets.

- **Construction of dykes and protective walls**: The construction of dykes and protective walls along the shores of Lake Tanganyika is a common measure adopted to prevent flooding. For example, in Uvira in the DEMOCRATIC REPUBLIC OF CONGO, the Red Cross worked with local authorities to build sandbag dykes and concrete protection walls along the most vulnerable areas. These structures have reduced flooding in populated areas, protecting thousands of homes.

- **Improving drainage systems**: Urban and rural drainage systems are being upgraded to effectively manage rainwater and reduce the risk of flooding. In Kigoma, Tanzania, projects have been launched to widen and strengthen existing drainage channels, using resistant materials and modern construction techniques. These improvements have reduced seasonal flooding and minimized damage to infrastructure.

Regional and community planning
Land use and community planning are essential to reduce the vulnerability of communities to flooding and rising water levels.

- **Community relocation**: In areas where the risk of flooding is particularly high, it is sometimes necessary to relocate communities to safer areas. For example, in the Mpulungu district of Zambia, several villages were relocated with the help of the government and international NGOs. Relocation sites were identified taking into account security criteria and access to essential resources, such as water and farmland.

- **Zoning and building regulations**: Zoning and building regulations are tightened to prevent construction in areas at high risk of flooding. In Tanzania, the government has put in place strict guidelines to prohibit the construction of new infrastructure in identified flood zones. These policies also encourage the renovation of existing buildings to make them more flood-resistant.

Farming practices adapted to climate change
Agricultural practices need to be adapted to cope with the impacts of climate change, including rising sea levels and variations in rainfall patterns.

- **Flood-resistant crops**: Farmers are encouraged to adopt crop varieties that are resistant to flooding and fluctuating water conditions. For example, submergent rice varieties, which can tolerate prolonged periods of submergence, have been introduced into farming areas around Lake Tanganyika. These varieties have shown greater resilience and higher yields under flooded conditions.

- **Water management techniques**: Water management techniques, such as drip irrigation and rainwater harvesting, are promoted to improve water use efficiency and reduce dependence on unreliable water sources. In Bujumbura, Burundi, pilot projects have demonstrated the effectiveness of these techniques, increasing agricultural yields and farmers' resilience to changing climatic conditions.

Awareness-raising and training
Raising awareness and training local communities are crucial to the effective implementation of adaptation strategies.

- **Community awareness programs**: Awareness programs are set up to inform communities about the risks associated with rising waters and the adaptation measures available. These programs use various means of communication, including local media, community workshops and school campaigns. For example, AFPDE in the DRC regularly organizes workshops to raise awareness of water resource management practices and adapted agricultural techniques.

- **Training in resistant construction techniques**: Local communities receive training in flood- and weather-resistant construction techniques. This training includes practical sessions on the use of local and sustainable materials, as well as demonstrations of elevated house construction. In Tanzania, training programs supported by the NGO Practical Action have enabled many families to strengthen their homes against the risk of flooding.

Emissions reduction and resource management policies
To mitigate the effects of climate change, it is essential to put in place policies to reduce greenhouse gas emissions and manage natural resources sustainably. To be effective, these policies must be integrated at local, national and international levels.

Reducing greenhouse gas emissions

Policies to reduce greenhouse gas (GHG) emissions are crucial to mitigating climate change and its impact on Lake Tanganyika.

- **Promoting renewable energies**: Governments and NGOs are encouraging the use of renewable energies to reduce dependence on fossil fuels. Projects to install solar panels, wind turbines and hydroelectric power stations are underway around Lake Tanganyika. In Zambia, for example, a solar farm project near Mpulungu was launched in 2021, providing electricity to over 10,000 homes and reducing C02 emissions.

- **Energy efficiency and reduced deforestation**: Initiatives to improve energy efficiency and reduce deforestation are being implemented to cut GHG emissions. Clean cooking programs, which promote the use of energy-efficient stoves, reduce firewood consumption and carbon emissions. In Tanzania, the "Clean Cookstove" initiative has distributed over 50,000 eco-friendly stoves to riverside communities, helping to reduce deforestation and emissions.

Sustainable management of natural resources
Sustainable management of natural resources is essential to preserve the Lake Tanganyika ecosystem and support the livelihoods of local communities.

- **Fisheries regulations and protection of biodiversity**: Governments implement strict regulations for fisheries management and protection of the lake's biodiversity. These regulations include fishing quotas, closed fishing seasons and the creation of marine protected areas. In the DEMOCRATIC REPUBLIC OF THE CONGO, the Ministry of Fisheries has established fish reserves in certain parts of the lake, prohibiting fishing to allow stocks to regenerate.

- **Integrated watershed management**: Integrated watershed management is adopted to preserve water quality and ecosystem health. This includes forest protection, wetland restoration and sustainable agricultural land management. The FAO's Integrated Watershed Management project in Tanzania, for example, aims to reduce soil erosion, improve water quality and promote sustainable agricultural practices.

Regional and international cooperation
Regional and international cooperation is crucial to the implementation of emission reduction and sustainable resource management policies.

- **Multilateral climate agreements**: The countries bordering Lake Tanganyika participate in multilateral climate agreements, such as the Paris Agreement, to commit to reducing their GHG emissions and adopting sustainable practices. These agreements facilitate cooperation and

sharing resources to achieve common emission reduction targets.

- **Partnerships with international institutions**: Partnerships with international institutions, such as the World Bank, the United Nations Development Programme (UNDP) and the Global Environment Facility (GEF), provide financial and technical support for sustainable management initiatives. For example, a partnership between the Burundian government and UNDP has financed reforestation and energy efficiency projects, helping to reduce emissions and protect ecosystems.

Conclusion

Adaptation and mitigation strategies in the face of rising water levels on Lake Tanganyika are crucial to minimizing negative impacts on local communities and the ecosystem. Adaptation measures, such as the construction of flood-resistant infrastructure, land-use planning and the adoption of adapted agricultural practices, enable communities to better prepare for climate change. At the same time, policies to reduce greenhouse gas emissions and manage natural resources sustainably are essential to mitigate the underlying causes of climate change and protect Lake Tanganyika's environment.

6.3 Importance of regional and international cooperation

Regional and international cooperation is essential to tackle the complex challenges posed by the rising waters of Lake Tanganyika. The transboundary nature of the lake means that actions taken by one country can have repercussions on other riparian countries. Consequently, a coordinated and collaborative approach is essential for sustainable resource management, flood risk reduction and the promotion of resilience to the impacts of climate change. This chapter looks in detail at regional programs and agreements, and the role of international organizations in managing Lake Tanganyika.

Regional programs and agreements

Regional programs and agreements play a crucial role in managing resources and protecting the Lake Tanganyika ecosystem. These initiatives promote collaboration between riparian countries, harmonize

environmental policies and facilitate the implementation of cross-border projects.

Lake Tanganyika Authority (ALT)
The Lake Tanganyika Authority (LTA) is an intergovernmental organization created to promote the sustainable management of Lake Tanganyika. ALT brings together the four countries bordering the lake - Tanzania, Burundi, the Democratic Republic of Congo (DRC) and Zambia - and coordinates their efforts to protect the lake's ecosystem and improve the livelihoods of local communities.

- **Regional biodiversity conservation program**: LALT has launched a regional biodiversity conservation program to protect endemic species and critical habitats in Lake Tanganyika. This program includes the creation of nature reserves, the monitoring of fish populations and the restoration of degraded habitats. For example, the Mahale Biodiversity Reserve in Tanzania has been established to protect endemic cichlid fish populations, and efforts to restore coral reefs have been undertaken in the area.

- **Integrated water resource management initiatives**: LALT coordinates integrated water resource management initiatives to improve water quality and watershed management. These initiatives include reforestation of degraded areas, pollution reduction and the promotion of sustainable agricultural practices. In 2020, a cross-border reforestation project was launched to restore forests around the lake and reduce soil erosion.

Environmental cooperation protocol
In 2008, the countries bordering Lake Tanganyika signed an environmental cooperation protocol under the aegis of the ALT. This protocol establishes a legal framework for the protection of the lake's environment and the sustainable management of its resources.

- **Harmonization of environmental policies**: The environmental cooperation protocol encourages the harmonization of environmental policies between riparian countries. Governments adopt common regulations for fisheries management, habitat protection and pollution reduction. For example, fishing quotas

have been established for certain fish species to prevent overfishing and allow stocks to regenerate.

- **Monitoring and evaluation mechanisms**: The protocol provides for monitoring and evaluation mechanisms to measure the impact of environmental initiatives and ensure their effectiveness. Joint monitoring teams collect data on water quality, fish populations and environmental changes in the lake. These data are shared between member countries to inform policy decisions and adapt management strategies.

Joint monitoring program
The joint monitoring program, coordinated by ALT, aims to collect data on the state of the Lake Tanganyika ecosystem and monitor the impacts of climate change and human activities.

- **Water quality monitoring**: The program includes regular water quality monitoring to detect pollution levels and changes in physico-chemical parameters. Monitoring stations are installed at several sites around the lake to measure nutrient levels, contaminants and water temperature. In 2022, data collected revealed an improvement in water quality in some areas thanks to pollution reduction efforts.

- **Fish stock monitoring**: The program monitors fish populations to assess the health of fisheries and detect overfishing trends. Periodic surveys are carried out to estimate fish biomass and identify species in decline. The results of these surveys are used to adjust fishing quotas and implement targeted conservation measures.

Cross-border conservation projects
Cross-border conservation projects are implemented to protect ecosystems and promote sustainable management of natural resources. These projects encourage cooperation between riparian countries and involve local communities in conservation activities.

- **Ruzizi delta conservation project**: The Ruzizi delta, located on the border between the DRC and Burundi, is an ecologically important wetland. A cross-border conservation project, supported by ALT and

funded by the United Nations Environment Programme (UNEP)$_z$ aims to protect this area by rehabilitating degraded habitats, building the capacity of local managers and involving communities in conservation activities. By 2021, the project has succeeded in restoring over 50 hectares of wetlands and increasing the population of certain species of aquatic birds.

- **Cross-border reforestation initiative**: riparian countries collaborate on cross-border reforestation projects to combat soil erosion and improve watershed resilience. A notable example is the reforestation project led by Tanzania and Zambia, which aims to reforest degraded areas along their shared border. By 2022, over 300,000 trees had been planted as part of this project, covering an area of 150 hectares.

The role of international organizations

International organizations play a crucial role in the management of Lake Tanganyika, providing financial, technical and logistical support for conservation and sustainable management initiatives. They also facilitate cooperation between riparian countries and promote the integration of environmental objectives into development policies.

Global Environment Facility (GEF)

The Global Environment Facility (GEF) is funding several conservation and sustainable management projects around Lake Tanganyika. The GEF supports initiatives aimed at protecting biodiversity, improving water resource management and strengthening the resilience of local communities to the impacts of climate change.

- **Sustainable Fisheries Management Project** : The GEF-funded Sustainable Fisheries Management Project aims to improve the management of fish stocks and promote sustainable fishing practices. The project includes training for local fishermen, setting fishing quotas and monitoring fish populations. Since the project began in 2018, catches of some fish species have shown signs of stabilization.

- **Biodiversity conservation projects**: FEM also funds biodiversity conservation projects targeting critical habitats and endangered species. For example, a coral reef conservation project in the southern part of Lake Tanganyika has restored

several degraded reef sites and strengthen the resilience of marine ecosystems.

United Nations Development PROGRAMME (UNDP)
The United Nations Development Programme (UNDP) supports sustainable development initiatives that integrate environmental and social objectives. UNDP works with local governments, NGOS and communities to promote sustainable practices and strengthen the resilience of vulnerable populations.

- **Climate resilience program**: UNDP has launched a climate resilience program to help communities around Lake Tanganyika adapt to the impacts of climate change. The program includes capacity-building activities, reforestation projects and sustainable water resource management initiatives. In 2020, the program supported the planting of over 200,000 trees and the construction of rainwater harvesting systems in several villages.

- **Community development initiatives**: UNDP also supports community development initiatives aimed at improving the livelihoods of local populations while protecting the environment. For example, a micro-credit project has been set up to finance sustainable income-generating activities such as beekeeping and aquaculture. This project has enabled many families to increase their income and reduce their dependence on natural resources.

Chapter 7: Conclusion

Lake Tanganyika, one of the oldest and largest lakes in the world, plays a central role in the lives of communities living along its shores in four countries: Tanzania, Burundi, the Democratic Republic of Congo (DRC) and Zambia. Throughout this book, we have explored various aspects of the rising waters of Lake Tanganyika, analyzing the causes, impacts and responses to this phenomenon.

We began with an overview of the geographical and ecological context of Lake Tanganyika, highlighting its importance for local communities and biodiversity. The lake is an essential source of fresh water, food and livelihoods for millions of people. However, it also faces significant challenges from rising water levels, a complex issue resulting from a number of factors, including climate change and human activities.

The second chapter traced the origin and formation of Lake Tanganyika, highlighting its geographical and geological evolution. We explored the lake's unique biodiversity, with a particular focus on endemic species and aquatic ecosystems. Understanding the lake's natural history is crucial to assessing the current and future impacts of environmental change.

In the third chapter we identified the main causes of rising water levels in Lake Tanganyika, focusing on global climate change and the impact of human activities such as deforestation and intensive agriculture. We also examined the natural variability of water levels and its role in amplifying current problems.

The fourth chapter detailed the environmental impacts of rising sea levels, particularly on aquatic fauna and flora and coastal ecosystems. Endemic species, already vulnerable, are particularly threatened by changes in their habitat. Coastal ecosystems are also undergoing significant transformations, affecting biodiversity and habitat health.

We explored the socio-economic repercussions of rising sea levels on riverside communities. This includes the displacement of populations, changes in lifestyles and economic activities, as well as risks to food security and freshwater resources. Local populations have to adapt to new realities, often with limited support.

The final chapter was devoted to responses and solutions to rising sea levels. We looked at local and international resource management initiatives, adaptation and mitigation strategies, and the importance of regional and international cooperation. Concrete examples of projects and collaborations were presented to illustrate the efforts underway and the challenges to be overcome.

Outlook for the future of Lake Tanganyika

The future of Lake Tanganyika will depend on the ability of the riparian countries and the international community to implement effective strategies to mitigate the impacts of climate change and protect this vital ecosystem. Several possible scenarios and future forecasts can be envisaged depending on the actions taken today.

Optimistic scenario

In an optimistic scenario, regional and international cooperation efforts are strengthened, with significant investments in biodiversity conservation, sustainable resource management and climate change adaptation. Reforestation, fisheries management and habitat protection projects are successful, improving the resilience of the Lake Tanganyika ecosystem and local communities. Greenhouse gas emissions are reduced thanks to rigorous environmental policies and the adoption of renewable energies, helping to stabilize the regional climate.

Pessimistic scenario

In a pessimistic scenario, conservation and sustainable management efforts are insufficient, and greenhouse gas emissions continue to rise. The impacts of climate change intensify, with rising sea levels and extreme climatic variations. Lake Tanganyika's ecosystems are undergoing irreversible degradation, leading to the disappearance of many endemic species. Riparian communities are facing massive displacement, loss of livelihoods and growing food insecurity. Tensions between riparian countries are rising, exacerbating conflicts and hampering regional cooperation.

Intermediate scenario

In an intermediate scenario, progress is being made, but in an uneven and fragmented way. Some countries and regions are implementing effective strategies, while others are lagging behind. Conservation and sustainable management initiatives are showing positive results, but challenges persist due to climate variations and anthropogenic pressures. Local communities continue to adapt, but with limited resources and insufficient support. Regional cooperation is progressing, but political and economic obstacles hinder large-scale concerted action.

7.3 Call to action for conservation and sustainable management
To ensure a sustainable future for Lake Tanganyika and its riparian communities, it is imperative to adopt a proactive and integrated approach, involving political decision-makers, NGOS and local communities. Here are some recommendations to guide future action:

Recommendations for policymakers
- **Strengthen environmental policies**: Governments must adopt and enforce strict environmental policies to protect Lake Tanganyika's ecosystems. This includes regulating fisheries, protecting critical habitats and reducing pollution.

- **Promoting renewable energies**: Policymakers need to encourage the adoption of renewable energies to reduce greenhouse gas emissions. Tax incentives and subsidies can be put in place to support solar, wind and hydroelectric power projects.

- **Investing in resilient infrastructure**: Investment in flood-resilient infrastructure, such as dykes and drainage systems, is essential to protect local communities. Governments must also promote sustainable, weather-resistant construction practices.

Recommendations for NGOs
- **Building local capacity**: NGOs must continue to provide training and resources to build the capacity of local communities to adapt to the impacts of climate change. This includes

training in sustainable farming techniques, water management and flood-resistant construction.

- **Promoting awareness and education**: Awareness campaigns and educational programs are crucial to informing local communities about the risks of rising sea levels and adaptation measures. NGOs should work with schools, local media and community leaders to disseminate this information.

- **Facilitating regional cooperation**: NGOs can play a key role in facilitating regional and international cooperation. They can act as a bridge between governments, international institutions and local communities to promote joint initiatives and cross-border projects.

Recommendations for local communities

- **Adopt sustainable practices**: Local communities should be encouraged to adopt sustainable practices in agriculture, fishing and the use of natural resources. This includes using flood-resistant crop varieties, reducing deforestation and managing fisheries responsibly.

- **Involvement in conservation initiatives**: The active participation of local communities in conservation initiatives is essential to their success. Local people must be involved in the planning and implementation of conservation projects, and their traditional knowledge must be integrated into management strategies.

- **Strengthening community solidarity**: Community solidarity is an invaluable asset in meeting the challenges of rising sea levels. Communities need to strengthen their mutual support networks, share resources and work together to develop locally adapted solutions.

Importance of ongoing commitment and awareness-raising

Continued commitment and awareness are essential to ensure the success of efforts to conserve and sustainably manage Lake Tanganyika. Political decision-makers, NGOs and local communities must maintain constant vigilance and a willingness to adapt to new challenges. Awareness of the importance of biodiversity conservation and sustainable resource management must be integrated into educational programs and communication campaigns.

The media play a crucial role in disseminating information and raising public awareness. Reports, documentaries and articles on the challenges and solutions associated with the rising waters of Lake Tanganyika can mobilize public opinion and encourage collective action.

Finally, scientific research and data collection are fundamental to understanding the environmental and socio-economic dynamics of Lake

Tanganyika. Researchers, academic institutions and conservation organizations need to work together to carry out in-depth studies.

Bibliography

1. Attn, S. R., et al. (2002). Effects of Land-use Change on Aquatic Invertebrate Diversity and Production in Lake Tanganyika, East Africa. Conservation Biology, 16(2), 493-503.

2. Allen, DJ., et al. (2018). Freshwater Ecoregions of the World: A New Map of Biogeographic units for Freshwater Biodiversity Conservation. BioScience, 68(5), 269-279.

3. Baratti, M., et al (2003). Molecular Phylogeny of the Genus Lamprologus (Teleostei: Cichlidae) and Implications for the Evolution of its Species Flock in Lake Tanganyika. Molecular Phylogenetics and Evolution, 27(1), 84-96.

4. Bootsma, H. A., & Hecky, R E. (2003). A Comparative Introduction to the Biology and Limnology of the African Great LakesJournal of Great Lakes Research, 29, 3-18.

5. Brander, L. M., et al (2007). The Economic value of Environmental Functions in the Okavango Delta, Botswana. Gland, Switzerland: IUCN.

6. Burgess, N. D., et al (2004). Terrestrial Ecoregions of Africa and Madagascar: A Conservation Assessment. Washington, D.C.: Island Press.

7. Chale, F. M. M. (2004). Pollution of Lake Tanganyikajournal of Great Lakes Research, 30(4), 465-476.

8. Cohen, A. s., et al. (1993). Environmental Change in Lake Tanganyika: The Sedimentary Record of One of the Most Ancient and Deepest Tropical Lakes. In D. G. George,J. GJones, P. Puncochar, C. S. Reynolds, & D. W. Sutcliffe (Eds.), Sediment Records of Biomass and Productivity (pp. 3-18). Dordrecht: Springer.

J . Cohen, A. S., et al. (2005). Ecological Consequences of Early Late Pleistocene Megadroughts in Tropical Africa. Proceedings of the National Academy of Sciences, 104(42), 16422-16427.

10. Coulter, G. W. (1991). Lake Tanganyika and Its Life. Oxford: Oxford University Press.

11. Eggermont, H., et al. (2010). Limnological and Ecological Responses to Environmental Change in East African Lakes: A Review. Freshwater

Biology, 55(12), 2418-2433.

12. Hecky, R. E., et al (1991). The Pelagic Ecosystem. In G. W. Coulter (Ed.), Lake Tanganyika and its Life (pp. 43-56). Oxford: Oxford University Press.

13. johnson, T. C., et al. (1996). Late Pleistocene Desiccation of Lake victoria and Rapid Evolution of Cichlid Fishes. Science, 273(5278), 1091-1093.

14. KoldingJ., & van Zwieten, P. A. M. (2006). The Tragedy of Our Legacy: How Do Global Management Discourses Affect Small Scale Fisheries in the South? Forum for Development Studies, 33 (1), 29-58.

15. Kullander, S. O. (2003). Family Cichlidae (Cichlids). In R. E. Reis, S. O. Kullander, & C.J. FerrarisJr. (Eds.), Checklist of the Freshwater Fishes of South and Central America (pp. 605-654). Porto Alegre: Edipucrs.

16. Langenberg, v. T., et al. (2003). Seasonal and Spatial variations in Phytoplankton Biomass and Primary Production in Lake Tanganyika. Freshwater Biology, 48(10), 1608-1623.

17. Lowe-McConnell, R. H. (2009). Fisheries and Cichlid Evolution in the African Great Lakes: Progress and Problems. Freshwater Reviews, 2(2), 131151.

18. McIntyre, P. B., et al. (2006). Historical Water Quality and Fishery Changes in Lake Tanganyika Inferred from Sediment Records. Freshwater Biology, 51 (5) ,J85-997.

19. Moore, M. v., et al. (1995). Zooplankton in the Open Waters of Lake Tanganyika: A Focal Point for Ecological and Evolutionary Investigation. Advances in Limnology, 45, 269-279.

20. Mugidde, R. (2001). Nutrient Status and Planktonic Nitrogen Fixation in Lake victoria, Africa. Limnology and Oceanography, 46(4), 815-826.

21. Olago, D. o., & Odada, E. O. (2007). Climate Change and the Western Rift valley Lakes in Kenya. In M. L. Parry, O. F. CanzianiJ. P. Palutikof, PJ. van der Linden, & C. E. Hanson (Eds.), Climate Change 2007: Impacts, Adaptation and vulnerability (pp. 423-467). Cambridge: Cambridge

University Press.

22. O'Reilly, C. M., et al (2003). Climate Change Decreases Aquatic Ecosystem Productivity of Lake Tanganyika, Africa. Nature, 424(6950), 766-768.

23. Ptmm, S. L., et al (1995). The Future of Biodiversity. Science, 269(5222), 347-350.

24. Plisnier, P. D., et al. (1992). Limnological Annual Cycle Inferred from Physical-Chemical Fluctuations at Three Stations of Lake Tanganyika. Hydrobiologia, 331(1), 53-71.

25. Reid, R. S., et al. (2000). Human Population Growth, Land use, and Food Supply in the Lake victoria Basin. In C. B. Barrett, F. Place, & A. A. Aboud (Eds.), Natural Resources Management in African Agriculture (pp. 45-56). Oxford: CABI Publishing.

26. Ribbink, AJ (1987). African Cichlid Lakes and Their Conservation: A Socio-economic Perspective. Environmental Biology of Fishes, 19(3), 241255.

27. Rusell,J. M., et al. (2007). Human Impacts, Environmental Change, and the Future of African Lakes. Environmental Science & Technology, 41(3), 755-760.

28. Sarvalaj, et al (2003). Trophic Structure of Lake Tanganyika: Carbon Flows in the Pelagic Food Web. Hydrobiologia, 500(1-3), 85-103.

29. Scheffer, M. (2009). Critical Transitions in Nature and Society. Princeton: Princeton university Press.

30. Seehausen, O. (2006). African Cichlid Fish: A Model System in Adaptive Radiation Research. Proceedings of the Royal Society B: Biological Sciences, 273(1597), 1987-1998.

31. Spigel, R. H., & Coulter, G. W. (1996). Comparison of Hydrology and Physical Limnology of the East African Great Lakes: Tanganyika, Malawi, victoria, Kivu, and Turkana (with Reference to Some North American Great Lakes). In T. CJohnson & E. O. odada (Eds.), The Limnology, Climatology, and Paleoclimatology of the East African Lakes (pp. 103-139). Amsterdam: Gordon and Breach.

32. Sturmbauer, C., et al. (2001). Evolutionary History of the Lake Tanganyika Cichlid Tribe Tropheini (Teleostei: Cichlidae) Derived from Mitochondrial DNA Sequences. Molecular Phylogenetics and Evolution, 22(3), 367-378.

33. Takahashi, T., et al. (2001). Geological and Biological Evolution of Lake Tanganyika: A Synthesis. In T. C. Johnson & E. O. Odada (Eds.), The Limnology, Climatology, and Paleoolimatology of the East African Lakes (pp. 677-698). Amsterdam: Gordon and Breach.

34. Thornton,j. A., et al. (1996). Lake Tanganyika: Experience and Lessons Learned Brief. In M. Munawar & R. E. Hecky (Eds.), The Great Lakes of the World (GLOW): Food-web, Health, and Integrity (pp. 109-115). Leiden: Backhuys.

35. Tierney, J. E., et al. (2010). Late Pleistocene Desiccation of Lake Tanganyika Inferred from Stable Isotope Analysis of sediment Cores. Quaternary science Reviews, 29 (7-8),

yes
I want morebooks!

Buy your books fast and straightforward online - at one of world's fastest growing online book stores! Environmentally sound due to Print-on-Demand technologies.

Buy your books online at
www.morebooks.shop

Kaufen Sie Ihre Bücher schnell und unkompliziert online – auf einer der am schnellsten wachsenden Buchhandelsplattformen weltweit! Dank Print-On-Demand umwelt- und ressourcenschonend produziert.

Bücher schneller online kaufen
www.morebooks.shop

Printed by Books on Demand GmbH, Norderstedt / Germany